YOUR KNOWLEDGE HAS VALUE

AF153488

- We will publish your bachelor's and master's thesis, essays and papers

- Your own eBook and book - sold worldwide in all relevant shops

- Earn money with each sale

Upload your text at www.GRIN.com and publish for free

On the disposition of the prime numbers. A matrix of all odd numbers

William Fidler

Bibliographic information published by the German National Library:

The German National Library lists this publication in the National Bibliography; detailed bibliographic data are available on the Internet at http://dnb.dnb.de.

ISBN: 9783346548702
This book is also available as an ebook.

On the disposition of the prime numbers

W M Fidler

Abstract

A mathematical structure is formed which consists of a matrix of all of the odd numbers. It then follows that this matrix contains all of the prime numbers with the exception of **2**. In conjunction with a prime number counting function denoted by $\pi(x)$ we can determine the number of primes in any row of the matrix and their location using the prime number search method developed, employing the prime number counting function data compiled from a number of sources by A V Kulsha [1] (and hereinafter referred to as 'Kulsha's data). The columns of the matrix are of infinite extent and in some there are either one or no prime numbers. In a column where there is a single prime number, this, in all instances appears in the first row. The matrix reveals that there are prime numbers which can never be twinned and that it is not possible for three consecutive odd numbers to be prime. Further, it is demonstrated, in some instances, even although the distribution of the primes in any given row is random, that it is possible to predict the next prime, and sometimes the prime immediately following, thus predicting a set of twins. Moreover, it is shown that it is possible to predict the gaps between some of the primes. The work, and, in particular the Discussion, contains numerous examples of the associated procedures. In the following Appendix it is demonstrated that we can determine the primality, or otherwise of any odd number, without examining in detail the number itself. In the course of the work it becomes plain that the utility of Kulsha's data, in the context presented here, may be increased considerably if it is condensed by confining it solely to the odd numbers.

It is emphasized that, in contradistinction to formulae which predict particular primes, for example, those of Mersenne and Sophie Germain, the analysis presented here makes no distinction between any of the 'types' of the prime numbers.

List of Contents

Introduction

The definition of what constitutes a prime number depends upon the mathematical interest of the definer, for it differs between number theorists, function theorists, algebraists and logicians. We choose here the best known definition, which is that of the number theorist and defines a prime number as that number which is divisible only by unity and itself. The prime numbers are all odd numbers with the exception of **2.**

The importance of the prime numbers cannot be overstated for they are the building blocks of all of the other numbers. Their study has occupied some of the best minds in Mathematics over millennia and the greatest mystery relates to their disposition amongst the odd numbers.

If the odd number line is laid out then the disposition of the prime numbers therein resembles the careless sprinkling of condiments on food, and attempts to explain this seemingly random disposition have met with only limited success.

It is shown in this work that the decomposition of the odd number line into the rows of a matrix together with the derivation of a prime number counting function which is specific to each row imposes a kind of order on the primes, to the extent that, in many instances 'next primes' may be predicted with confidence, together with some of the gaps between the primes and, in some cases, the presence of prime twins.

Analysis

It was shown in previous work [1a], by the author that the whole odd number line could be decomposed and reassembled to form a matrix having an infinite number of rows and fifteen columns. Parts of the matrix are shown immediately below and later. The first part is in tabulated form whilst the second, for reasons of space, is shown as an array.

To facilitate the argument presented here the prime numbers have been determined a priori using a prime number calculator and appear in boldface and underlined.

It should be noted that the numerical spacing between adjacent rows is **30**.The numbers at the bottom of each presentation are the column numbers and it may be seen immediately, with the exception of the first row, that there are columns in which there are no prime numbers. In the search for prime numbers this is of immediate benefit.

3	**5**	**7**	9	**11**	**13**	15	**17**	**19**	21	**23**	25	27	**29**	**31**
33	35	**37**	39	**41**	**43**	45	**47**	49	51	**53**	55	57	**59**	**61**
63	65	**67**	69	**71**	**73**	75	77	**79**	81	**83**	85	87	**89**	91
93	95	**97**	99	**101**	**103**	105	**107**	**109**	111	**113**	115	117	119	121
123	125	**127**	129	**131**	133	135	**137**	**139**	141	143	145	147	**149**	**151**
153	155	**157**	159	161	**163**	165	**167**	169	171	**173**	175	177	**179**	**181**
183	185	187	189	**191**	**193**	195	**197**	**199**	201	203	205	207	209	**211**
213	215	217	219	221	**223**	225	**227**	**229**	231	**233**	235	237	**239**	**241**
243	245	247	249	**251**	253	255	**257**	259	261	**263**	265	267	**269**	**271**
273	275	**277**	279	**281**	**283**	285	287	289	291	**293**	295	297	299	301
1	2	3	4	5	6	7	8	9	10	11	12	13	14	15

Table 1

Now, we may use a prime number counting function to determine the number of primes in any given row, and by interrogation of the numbers in columns **3, 5, 6, 8, 9, 11, 14, 15**, obtain the distribution of the primes in that row. However, this involves showing the primality, or otherwise of any given number in that row. We show in the Appendix, p20, that we can construct the matrix, identifying all of the prime and composite numbers without actually examining the numbers themselves.

Indeed, we can, in certain instances attain what is considered by many number theorists to be the Holy Grail of Number Theory by predicting, with absolute certainty the next prime number in a given, and random sequence. It should be emphasized here that we are not speaking of what could be termed 'formulaic primes' such as those of Mersenne and Sophie

Germain, but rather, those random sequences of primes which arise in the list of the odd numbers.

Let us consider Table 1 above.

If we proceed along the first row and thence to the number in the second column of row **2** we arrive at the number **35**. It is easy to see that all of the prime numbers below **35** are contained in row **1**. The reason for terminating at **35** is that it is the last number in this sequence of prime numbers at which we can be confident that this last number is not prime. Whilst the next number in the sequence may not be prime we cannot be sure of this, and so we terminate at **35**.

From Kulsha's data we find that the prime number counting function, $\pi(x)$ corresponding to **35** is **11**.

We see in row **1** that there are **10** prime numbers and this may be explained by the fact that $\pi(x)$ also includes the first prime number, viz. **2**, and this does not appear in the matrix.

Hence, $\pi(35) - \pi(2) = 11 - 1 = 10$, the number of primes in the first row.

In this instance it is seen that we may predict the last prime number in this set, for there are only certain columns that may contain a prime number, and having filled the columns **1,2, 3, 5, 6, 8, 9, 11, 14** it follows that the number in the only remaining column, i.e. **15** must be prime.

Of course, this is a trivial example but the same reasoning may be applied to any row, for example, consider row **5**, where the terminating number in the type of sequence already established is **125**. Now, from Table 1, $\pi(125) = 30$, and $\pi(95) = 24$. Hence the number of primes in row 4 is $30 - 24 = 6$, and this is confirmed by inspection. We can then investigate the columns which could contain prime numbers to determine the location of the primes in this row. It is reiterated that since we are considering the difference between the magnitude of two prime number counting functions then $\pi(2)$ disappears from the calculation.

All of the above could be considered fortuitous and so we repeat the calculation for another row.

Consider row **9**. The terminating number here, in the sense of the foregoing is **245**, hence, the number of primes in row **8** is $\pi(245) - \pi(215) = 53 - 47 = 6$. Inspection of row **8** shows that it indeed contains **6** prime numbers; starting from the left we encounter the first prime in column **6**. Since columns **7, 10, 12**, and **13** cannot contain primes it follows that the numbers in columns **8, 9, 11**, and **14, 15** are prime. Thus, at a stroke we have predicted an isolated prime and two sets of twins.

The reader is invited to predict the number of primes, and their disposition in any row of interest in Table 1 using the values extracted from Kulsha's data in Table 2, and the methodology presented above.

Table 2

x	$\pi(x)$
35	11
65	18
95	24
125	30
155	36
185	42
215	47
245	53
275	58
305	62

It will have become obvious that the argument of the prime number counting function used in the above has the form, $5[1 + 6(r - 1)]$, where **r** is the row number and will yield the 'bounding numbers' (e.g **245** and **215**) such that the number of primes in row **r – 1** (i.e. row **8**) may be determined. Equally, we may add **30** to **245** to yield the pair (**275** and **245**) and hence determine the number of primes in row **9**.

The general location of any number in the matrix is obtained by the following procedure, which is best illustrated by an example.

Consider the number **219**, chosen at random. Now, if we subtract **9** from this we get **210**, which is divisible by **30**. The row number is then **210/30 + 1 = 8**, and the column number is identified by the number subtracted from **219**, i.e. **9**. It then follows that the column number is **4**.

It may be noted, for the matrix, that any of the numbers, **N**, therein may be obtained from the simple formula

$$N = 2[\,c + (r - 1)\hat{c}\,] + 1 \text{------- (a).}$$

Here, **c** is the column number, **r** the row number and $\hat{c}$ the maximum column number.

The array following is a part of the matrix, chosen at random. Using the procedure previously explained then, the first row number is calculated from **(2013 – 3)/30 + 1 = 68** and the column number is **1**, for we have subtracted **3** from the number **2013** to determine the row number.

2013 2015 **2017** 2019 2021 2023 2025 **2027** **2029** 2031 2033 2035 2037 **2039** 2041 -- (68)

2043 2045 2047 2049 2051 **2053** 2055 2057 2059 2061 **2063** 2065 2067 **2069** 2071 --(69)

2073 2075 2077 2079 **2081** **2083** 2085 **2087** **2089** 2091 2093 2095 2097 **2099** 2101 --(70)

2103 2105 2107 2109 **2111** **2113** 2115 2117 2119 2121 2123 2125 2127 **2129** **2131** --(71)

2133 2135 **2137** 2139 **2141** **2143** 2145 2147 2149 2151 **2153** 2155 2157 2159 **2161** --(72)

2163 2165 2167 2169 2171 2173 2175 2177 **2179** 2181 2183 2185 2187 2189 2191 --(73)

2193 2195 2197 2199 2201 **2203** 2205 **2207** 2209 2211 **2213** 2215 2217 2219 **2221** --(74)

2223 2225 2227 2229 2231 2233 2235 **2237** **2239** 2241 **2243** 2245 2247 2249 **2251** --(75)

2253 2255 2257 2259 2261 2263 2265 **2267** **2269** 2271 **2273** 2275 2277 2279 **2281** --(76)

2283 2285 **2287** 2289 2291 **2293** 2295 **2297** 2299 2301 2303 2305 2307 **2309** **2311** --(77)

2313 2315 2317 2319 2321 2323 2325 2327 2329 2331 **2333** 2335 2337 **2339** **2341** --(78)

2343 2345 **2347** 2349 **2351** 2353 2355 **2357** 2359 2361 2363 2365 2367 2369 **2371** --(79)

2373 2375 **2377** 2379 **2381** **2383** 2385 2387 **2389** 2391 **2393** 2395 2397 **2399** 2401 --(80)

2403 2405 2407 2409 **2411** 2413 2415 **2417** 2419 2421 **2423** 2425 2427 2429 2431 --(81)

2433 2435 **2437** 2439 **2441** 2443 2445 **2447** 2449 2451 2453 2455 2457 **2459** 2461 --(82)

2463 2465 **2467** 2469 2471 **2473** 2475 **2477** 2479 2481 2483 2485 2487 2489 2491 --(83)

2493 2495 2497 2499 2501 **2503** 2505 2507 2509 2511 2513 2515 2517 2519 **2521** --(84)

2523 2525 2527 2529 **2531** 2533 2535 2537 **2539** 2541 **2543** 2545 2547 **2549** **2551** --(85)

2553 2555 **2557** 2559 2561 2563 2565 2567 2569 2571 2573 2575 2577 **2579** 2581 --(86)

2583 2585 2587 2589 **2591** **2593** 2595 2597 2599 2601 2603 2605 2607 **2609** 2611 --(87)

1 2 3 4 5 6 7 8 9 10 11 12 13 14 15

As before, we present in Table 3 the relevant selection of data from Kulsha's prodigious work. We note Zagier's [2] statement that the extensive tables of the prime number counting function represent 'thousands of hours of dreary calculation' (sic), although this may be more of a comment on the status of computing circa 1975.

Choosing row **78** at random we can calculate that the number of primes in row77 is:

$$\pi(2315) - \pi(2285) = 344 - 339 = 5.$$

In addition, starting from the left we can 'fill' the appropriate columns and without further ado state that the number 2311 is prime by observing that this is the only remaining column which can contain the remaining prime number..

Indeed, since there is no stricture on the 'direction' in which the primes are uncovered then, equally, we could have started at column 15 and proceeded towards the left; the number in column 5 would have been investigated for primality and been shown to be not prime; the number in column 4 cannot be prime for it is a member of a 'non-prime column' and so the only remaining location at which the last prime could be located is column 3, hence, the number 2287 is prime.

We may apply this reasoning to any row.

Table 3

x	$\pi(x)$
2015	305
2045	309
2075	312
2105	317
2135	321
2165	326
2195	327
2225	331
2255	335
2285	339
2315	344
2345	347
2375	351
2405	357
2435	360
2465	364
2495	367
2525	369
2555	374
2585	376
2615	379

In the determining of the numbers of primes in any given row of the matrix the prime number counting function need only be determined three times in every one hundred of the natural numbers. The success of the foregoing depends only upon access to Kulsha's data set for we can always determine the' bounding numbers' of the range of numbers in which the primes of a given row exist.

Formulae for the prediction of the number of primes in any row of the matrix

Riemann [4] proved that the prime number counting function, $\pi(x)$ is given exactly by:

$$\pi(x) = R(x) - \sum_\rho R(x^\rho) \text{ ----------- (1).}$$

Here, $R(x) = \sum_{n=1}^{\infty} \frac{\mu(n)}{n} Li(x^{1/n})$, where $\mu(n)$ is the Möbius function, and **Li** denotes the logarithmic integral; in honour of Riemann the function $\mathbf{R(x)}$ is called the Riemann prime number counting function.

If the trivial zeros of the zeta function are collected and the sum taken only over the non-trivial zeros, ρ then $\pi(x)$ may be written:

$$\pi(x) = R(x) - \sum_\rho R(x^\rho) - \frac{1}{\ln x} + \frac{1}{\pi} \tan^{-1} \frac{\pi}{\ln x} \text{ --------- (2).}$$

Further, as stated by Zagier [3], $\mathbf{R(x)}$ is an entire function of **ln x** and was proved by Hardy [4], to be equivalent to the rapidly-converging series derived by Gram [5] in 1884, i.e.

$$R(x) = 1 + \sum_{k=1}^{\infty} \frac{(\ln x)^k}{k\,k!\,\zeta(k+1)} \text{ -------- (3).}$$

Here, $\zeta(k+1)$ is the zeta function. According to Zagier [3], $\mathbf{R(x)}$ is an exceptionally good approximation to $\pi(x)$.

This is exemplified by the data which is shown in the following table, which is given in [3] without attribution.

x	$\pi(x)$	R(x)
100,000,000	5,761,455	5,761,552
200,000,000	11,078,937	11,079,090
300,000,000	16,252,325	16,252,355
400,000,000	21,336,326	21,336,185
500,000,000	26,355,867	26,355,517
1,000,000,000	50,847,534	50,847,455

The agreement between the counted number of primes, $\pi(x)$ and $\mathbf{R(x)}$ is truly astonishing.

We now use equation (3) to derive a pair of equations which will, in principle, predict the number of primes in any given row of the matrix.

A selection of rows of Table1 is now used to illustrate a procedure whereby the arguments of the functions referred to previously are determined.

Rows 5, 6, and 7 are shown below.

123	125	**127**	129	**131**	133	135	**137**	**139**	141	143	145	147	**149**	**151**
153	155	**157**	159	161	**163**	165	**167**	169	171	**173**	175	177	**179**	**181**
183	185	187	189	**191**	**193**	195	**197**	**199**	201	203	205	207	209	**211**

In row **7** the number preceding the first prime is **189** and the number **153** is preceded by the last prime in row **5**. It then follows that the number range **153---189** contains all the primes in row **6**. Hence, the number of primes in row 6 is $\pi(189) - \pi(153) = 42 - 36 = 6$. For clarity we now repeat this process for rows **4**, **5**, and **6**.

93	95	**97**	99	**101**	**103**	105	**107**	**109**	111	**113**	115	117	119	121
123	125	**127**	129	**131**	133	135	**137**	**139**	141	143	145	147	**149**	**151**
153	155	**157**	159	161	**163**	165	**167**	169	171	**173**	175	177	**179**	**181**

The number range here is **115----155.** Hence the number of primes in row **5** is:

$$\pi(155) - \pi(115) = 36 - 30 = 6.$$

In both examples we have used Kulsha's data to determine the number of primes in a particular row.

Whilst we have access to Kulsha's data then we may determine the number of primes in a row of interest without recourse to any equation..

If, on the other hand we wish our procedure to be self-contained, in the sense that it does not rely on an external source of data then, amongst other things we must first consider the definition of a prime number counting function.

Such a function is defined as some process which will yield the number of prime numbers in a range of numbers, up to and including the limit of the range.

If we denote the row of interest by **r**, then using equation (a) we may now, in the light of the foregoing, generalise the number range by the result of that equation, i.e.
$2\{ c^- + ĉ (r^- - 1)] + 1$ --------- $2[c^+ + ĉ (r^+ - 1)] + 1$, where the superscripts '-' and '+' denote the lower and upper bounds of the number range, and the numbers generated are the last and first primes in row '-' and row ' + ', respectively.

For brevity we write, $F = 2[c^+ + ĉ (r^+ - 1)] + 1$, and, $G = 2\{ c^- + ĉ (r^- - 1)] + 1.$

We now present an argument, based upon tabular data to justify the choice of the numbers bounding the range of interest, which will show that the form of the equation which will yield the number of primes in any row is not the simple subtraction, **R(F) – R(G)**.

We denote the numerator and denominator of equation (3) by **N** and **D**, respectively. Let **x** = **97**.

k	$D = kk!\,\zeta(k+1)$	$N = [\ln 97]^k$	N/D
10	36305926	4014547	0.11058
9	3269167.9	877552.1	0.2684
8	323207.85	191826.8	0.59351
7	35423.847	41932	1.18372
6	4356.08	9166.045	2.1042
5	610.406	2003.634	3.28408
4	99.545	437.98	4.39982
3	19.482	95.74	4.91425
2	4.80818	20.928	4.3526
1	1.644932	4.5747	2.78109
			$\Sigma = 23.99225$

Now, **R(97) = 25**, and from equation (3) and the above table we obtain: **R(97) = 1 +
23.99225 = 24.99225.**

From this we see that for the number **97,** equation (3) converges very rapidly.

Let us now consider **x = 101**.

k	$D = kk!\,\zeta(k+1)$	$N = [\ln 101]^k$	N/D
10	36305926	4383595	0.12074
9	3269167.9	949833.2	0.2967
8	323207.85	205809	0.63677
7	35423.847	44594.49	1.25888
6	4356.08	9662.695	2.2182
5	610.406	2093.704	3.43
4	99.545	453.6617	4.55735
3	19.482	98.299	5.04563
2	4.80818	21.2993	4.4298
1	1.644932	4.615121	2.80566
			$\Sigma = 24.7997$

Now, **R(101) = 26**, and from the above table and equation (3) we obtain: **R(101) = 1 +
24.7997 = 25.7997.**

Hence we see that equation (3) converges at a different rate for **x = 97** and **x = 101.**

It would then be a mistake to combine equation (3) in the following manner:

$$P_r = \sum_{k=1}^{\infty} {}^{1}\!/_{k\,k!}\;\zeta(k+1)\;\left\{[\ln F]^k - [\ln G]^k\right\} \text{------------- (4).}$$

Further, since we require the number of primes within a range bounded by two prime numbers, we must subtract unity from the following combination, for we have shown that there are no prime numbers <u>between</u> **97** and **101** and yet **R(101) – R(97) = 1**. Hence,

$$P_r = \sum_{k=1}^{\infty} \frac{(\ln F)^k}{k\,k!\;\zeta(k+1)} - [1 + \sum_{k=1}^{\infty} \frac{(\ln G)^k}{k\,k!\;\zeta(k+1)}] \text{---------- (5).}$$

It should be borne in mind that there can only be, at most, **8** prime numbers in any row of the matrix and so the value of the outcome of this equation can never exceed, **8**.

Given Zagier's statement regarding the accuracy of equation (3) and comparing the data shown in his accompanying table, it is then a reasonable expectation that a form of Riemann prime number counting function for any given row of the matrix, as represented by equations (3) and (5) , will be endowed with the same characteristic.

Since only the magnitude of the numerator of the above equation changes for each **x** then the calculation of the final result may be facilitated by constructing a library of the denominator terms, $k\,k!\;\zeta(k+1)$.

It may readily be argued that the forms for **F** and **G** require that we must perform additional work in determining the first and last primes in the rows adjacent to the row of interest. It is considered far more efficient if we only use a pair of a priori fixed numbers. This is the case if we revert to the initial form of the argument described on p7.

Hence, we set $F = 5[1 + 6r]$ and, $G = 5[1 + 6(r - 1)]$, and we rewrite equation (5) as:

$$P_r = \sum_{k=1}^{\infty} \frac{(\ln F)^k}{k\,k!\;\zeta(k+1)} - \sum_{j=1}^{\infty} \frac{(\ln G)^j}{j\,j!\;\zeta(j+1)} \text{----------(6).}$$

In view of the difference, as shown before between the results of the summation for **x = 101** and **x = 97** over the same range, we write the enumerators of equation (6) as shown to reflect that this difference may (probably will) occur in other situations. This is considered prudent, since it should be regarded as a maxim that nothing should be taken for granted when dealing with prime numbers.

There is a caveat that must be noted regarding equations (5) and (6). It is drawn to the attention of the reader that the data presented in Zagier's table on p10 show that, whilst exceptionally good, the agreement between $\pi(x)$ and **R(x)** is not perfect and the equations developed are sensitive to this discrepancy. For example, row **73** of the array on p8 contains only one prime, whilst equations (5) and (6) predict three primes if summed to only ten terms. This may be an instance of the disparity between Kulsha's counted data and the predictions from the Gram series representing Riemann's prime number counting function. Hence, these equations should be used with caution if the number of summations is small. Indeed, it is shown in [6] that there is a relatively large disparity between the Riemann Prime Counting function and Gram's series if insufficient terms are employed in the summation. However, since there can only be 8 primes in any row of the matrix after the first row, then the effort

13

involved in determining the location of the primes is not particularly great. Indeed, a systematic investigation of the rows of the matrix will reveal the exact locations where the predictions of the equations are not correct.

Further, in view of the magnitude of the term k*k! in the denominator of Gram's series it is of the utmost importance that the magnitude of the zeta function be determined to great precision. Unless this is done then the denominator will always be smaller than it should be, leading to an overestimate of the number of primes

Using a series calculator the following values were found for the number of primes in the range of row **73**. The calculator would only permit summation of up to **340** steps and, yields the following results:

2195-----327.567 primes 2203---328.8499 primes

2165-----323.7254 primes 2161---323.201 primes

The differences of **3.84** and **5.6489 – 1 = 4.6489** represent the number of primes in row **73** using equations (6) and (5), respectively.

These discrepancies are consistent with taking a small, relatively speaking, number of steps in the summations associated with the Riemann prime number counting function and Gram's series. Indeed, in [6] it is shown that close agreement between these two functions is only attained when step numbers of 10^4, 10^5 and 10^6 are employed, and such hugely exceeds the capacity of any series calculator to which the author has access.

Notwithstanding, the table on p10 shows that Riemann's prime counting function always overestimates the number of primes in the number ranges shown, and, it is offered, purely as a speculation, that we may have stumbled over a region where part of the overestimation occurs.

A local prime number density

The prime number density, ρ is usually taken to be the ratio of the number of prime numbers in any number range, **x** divided by that range.

This is an exceptionally blunt instrument for it may obscure the presence of a set of prime numbers within the range and which may be of interest.

The derivation of the local prime number counting function redresses this, for, it is shown clearly in, for example Table1 that there are only eight places in any row of the matrix which can be occupied by a prime number.

Hence, we may define a local prime number density, ρ_r as the number of primes in a given row divided by **8.**

14

i.e. $\rho_r = \frac{Pr}{8}$ -------------------- (7).

It then follows that throughout the whole of the matrix, the local prime number density as defined above can only vary between zero and unity.

Further, we may, in conjunction with Kulsha's data determine the local prime number density in the intervals: $2[1 + \hat{c}\,(r-1)] + 1$ ----------------- $2\hat{c}r + 1$, throughout the whole of the range of the odd numbers, without any knowledge of the prime numbers themselves.

Discussion

Provided that we can calculate the prime number counting function to any desired level then we may, in principle, uncover the distribution of the primes irrespective of the magnitudes of the numbers involved. For any row of the matrix we need only investigate, by any chosen method, the primality of the numbers in the appropriate columns. It is known that the density of the prime numbers decreases with increasing magnitude of the numbers and, it is possible that, in the case of very large numbers there may be rows which are devoid of primes, although this assertion would require an in-depth investigation using equations (5) or (6), where an outcome close to zero might be interpreted as the absence of a prime.

It was noted in [1] that the largest currently-known prime number is a Mersenne prime, which we denote by $\widehat{M}$, was discovered in December 2018 and is given by the formula: $2^{82589933} - 1$. To base 10 the number has **24,862,048** digits. The last two digits are **9** and **1**, and this is sufficient, under the following conditions, to determine the number of the column in which the prime resides.

We may only subtract either **11** or **31** from **91**, but we cannot subtract **21**, for the prime cannot be located in column **10** since there are no prime numbers therein.

We require someone with sufficient computational fortitude to sum, starting from the left, **24,862,046** digits of $\widehat{M}$ (note that we have excluded the last two numbers, **9** and **1**) then, denoting this sum by Σ, the prime will be found in column **5** if $(\Sigma + 8)$ is divisible by 3, or in column **15** if $(\Sigma + 6)$ is divisible by 3. The corresponding row number is given by either of $\frac{(\widehat{M} - 11)}{30} + 1$ or $\frac{(\widehat{M} - 31)}{30} + 1$.

Using exactly the same procedure described previously we can now determine whether or not the Mersenne prime described is the only prime in the row given by one of the row numbers above. This, of course will only be possible if we are in possession of accurate prime number counting function data up to at least the magnitude of $\widehat{M}$ or we possess equations which can, in principle, determine the number of prime numbers in any row of the matrix, or, at least eliminate these numbers which cannot be prime.

Whilst we have no intention of examining further this largest prime number we will investigate the Mersenne prime derived by Lucas in 1876. Denoting this prime number by $\boldsymbol{M_L}$, we have $\boldsymbol{M_L} = \mathbf{2^{127} - 1 = 170141183460469231731687303715884105727}$.

Now, we cannot subtract **27** from the above number for this would locate the prime number in the column where there are no primes. We can only subtract either **7** or **17.**

Now, in this instance, $\boldsymbol{\Sigma} = \mathbf{145 - (2 + 7) = 136}$. If we subtract **7** from $\boldsymbol{M_L}$ then the sum of the digits is now **136** + **2** = **138**, which is divisible by **3.** Subtracting **17** from $\boldsymbol{M_L}$ gives the sum of the digits as **136** + **1** = **137,** which is not divisible by **3**. Hence we conclude that the Lucas prime lies in column **3**.

The row number where this prime is located is given by:

$\mathbf{(170141183460469231731687303715884105720)}$ /**3** + **1** =

56713727820156410577229101238628035240 + **1**.

Without equation (5) the number of primes in this row is given by the somewhat daunting:

$\pi\mathbf{(170141183460469231731687303715884105755)}$ –
$\pi\mathbf{(170141183460469231731687303715884105725)}$, provided that such data exists.

The reader's attention is drawn to the last three digits of each of the above numbers. The numbers associated therewith are those in the row of interest in column **3** and the row immediately following it, respectively (i.e. the 'bounding numbers'). Moreover, it is easy to see that the numbers $M_L \pm 2$ cannot be prime.

With the exception of the sequence of prime numbers **2, 3, 5, 7** it is seen by inspection of any of the sections of the matrix that it is not possible for three consecutive numbers to be prime numbers. This is consonant with the observation that throughout the list of the odd numbers there are, at regular intervals, 'cage numbers' which are all divisible by 3 and encompass two other numbers which may, or may not, be prime.

If we examine the matrix it is seen that a prime number in column **3** is always isolated, in the sense that it can never be twinned for there is an intervening 'non-prime column' between it and the next possible prime. Hence, the Lucas prime is a member of that group of primes.

In the discussion of the gaps between the primes the numbers **113** and **127** are frequently used to illustrate how some gaps may be quite large, for no apparent reason. The matrix representation of the odd numbers gives an explanation of this. The number **113** lies in row **4** of the matrix and the prime counting function for this row shows that there are **6** primes therein.

Starting from the left, by examining each column in which a prime number may exist we determine that the number in each of columns **3, 5, 6, 8,** and **9** is prime. If we show further (and we do) that the number, **113** is prime then there can be no other primes in that row. This in itself is useful for it shows that the numbers in the other columns which can contain prime numbers (i.e. columns **14** and **15**) are not prime. In order to reach the next column which can contain a prime we must advance to the next row where, the first such column is found to be number **3.** The number occupying this location is **127,** which we can show is prime ---hence the extent of the gap.

An even more extreme example arises when we consider the gap between the number **2179** in row **73** and the next prime. Using the prime number function counting data we show that this is the only prime number in that row. The prime number counting function data shows that there are **4** prime numbers in row **74.** If we begin our search for the prime numbers in this row, starting this time from the right, we find that we have exhausted the number of primes that can exist in this row at column **6.** The number in this column is **2203,** hence we have an even greater spacing, viz. **2179** ----- **2203** here, than in the previous example.

Another strategy arises if we consider row **80.** The prime number counting function shows that there are **6** primes in this row. If we start from the right identifying primes as we move to the left, then we determine that after **2389** there is one column which can contain a prime — and we eliminate this by showing that the number **2387** in that column is composite. The next column cannot contain primes and since we have accounted for **3** of the primes, then there are **3** remaining primes to distribute between **4** columns, but column **4** contains no primes, hence it follows that the next two numbers are prime (and twins), whilst the number **2377** in column **3** is an isolated prime in the sense described earlier.

The author does not possess the computing facilities to pursue the method of solution of equations (5) and (6) for large numbers. However, the following example will illustrate that, providing data such as that of Kulsha is available, we may roam throughout the whole range of the odd numbers. Further, if we revert to the form of the arguments for **F** and **G** as explained on p14, then there is no need to rely upon any external data, but what is then required is a Gram series calculator capable of dealing with very large numbers, for the factor k*k! for even moderate k becomes very large.

At random, we guess an odd number, **15077** (which happens to be prime, but this is of no importance)

The location of this number in the matrix uses the same approach as shown previously, i.e.

15077 – 7 = 15070, which is not divisible by **30; 15077 – 17 = 15060** which is divisible by **30** and hence the row where the original number is located is **15060/30 + 1 = 503;** since we have subtracted **17** from the original number then the column number is **8.**

We now use this information to construct the part of the matrix which contains rows **502, 503** and **504** .

Some of the numbers in the following table are of no interest in the procedure and are simply indicated by '--'.

1	2	3	4	5	6	7	8	9	10	11	12	13	14	15
15033	--	--	--	--	--	--	--	--	--	--	--	--	--	**15061**
15063	15065	--	--	--	**15073**	--	**15077**	--	--	**15083**	--	--	--	**15091**
15093	15095	--	--	**15101**	--	--	--	--	--	--	--	--	--	--

The primes are shown in boldface and underlined.

The first prime in row **504** is **15101**, whilst the last prime in row **502** is **15061**.

Now, $\pi(15101) = 1764, and \pi(15061) = 1759$, the difference being **5**. However, there are only **4** primes in row **503**. The discrepancy is that explained previously and is resolved by noting that we require the difference between the two bounding primes. Hence we must subtract unity from the first result to give the numbers of primes in row **503** to be **4**.

If we take the bounding numbers to be **15095** and **15065**, then $\pi(15095) - \pi(15065) = 1763 - 1759 = 4$.

The 'prediction of the next prime' follows from enumerating the primes in row **503**, starting from the left; for, having determined the location of three of the primes and shown that the numbers in columns **12, 13** and **14** are composite, then the only remaining column must contain the remaining prime viz. **15091**. Further, we may determine the gap between the last prime in row **502** and the first prime in row **503** by starting from the extreme right of row **503** and moving to the left until we have exhausted the predicted number of primes. The gap **is** then **15061-----15073**. This is simply another version of that described earlier for determining a gap.

We may obtain a variation of this procedure by asking the question 'how many primes are there in row **1500**'.

We may obtain the upper 'range number' by $2[2 + 15*1500] +1 = 45005$, and hence the lower range number is $45005 - 30 = 44975$. Now, from Kulsha's data $\pi(45005) = 4675$.

whilst $\pi(44975) = 4673$. Hence there are **2** primes in row **1500**. Further, the local prime number density is then, **2/8**.

Finally, row **4** of Table1 is particularly interesting, for having determined that there are **6** primes therein, then, starting from the extreme right we can eliminate all of the numbers in columns **12, 13, 14,** and **15** for none can be shown to be prime. The number in column **11** is prime. We now proceed to column **1** and move rightwards. It is shown that the number in column **3** is prime. Now, columns **4, 7,** and **10** cannot contain primes. It then follows that the numbers in columns **5,6** and **8,9** must be prime. Thus, as in the example shown on p6, we have identified a pair of twin primes---without actually investigating them for primality..

We leave it to the interested reader to devise other strategies in a similar vein.

To give a flavour of the magnitude of numbers associated with work of this nature, it is quoted on p 248 of [7] that: $\pi(10^{20}) = 2220819602560918840$.

There is no merit in extending the number of columns by attaching consecutive rows to each other. Indeed, the matrix developed here is the simplest structure of its type---and the easiest to interrogate.

It is considered that the methods outlined here may be a contribution of significance to that part of number theory dealing with the disposition of the prime numbers.

W M Fidler, June 2021.

We now show that we can construct the matrix by systematically determining the nature of the numbers in the columns which can contain primes.

We reproduce Table1 for the purpose of illustration. The reader is asked to imagine that the only primes that have been determined are those in the first row of the matrix, and that, in addition, from row 2 onwards the only columns which can contain primes are 3, 5, 6, 8, 9, 11, 14, and 15.

3	5	7	9	11	13	15	17	19	21	23	25	27	29	31
33	35	37	39	41	43	45	47	49	51	53	55	57	59	61
63	65	67	69	71	73	75	77	79	81	83	85	87	89	91
93	95	97	99	101	103	105	107	109	111	113	115	117	119	121
123	125	127	129	131	133	135	137	139	141	143	145	147	149	151
153	155	157	159	161	163	165	167	169	171	173	175	177	179	181
183	185	187	189	191	193	195	197	199	201	203	205	207	209	211
213	215	217	219	221	223	225	227	229	231	233	235	237	239	241
243	245	247	249	251	253	255	257	259	261	263	265	267	269	271
273	275	277	279	281	283	285	287	289	291	293	295	297	299	301
1	2	3	4	5	6	7	8	9	10	11	12	13	14	15

Table 1

We may now, using only Kulsha's data determine the nature of all of the numbers in this part of the matrix, without examining, in detail the numbers themselves. We focus attention on column 3, for the approach to determining the disposition of the primes in this column is an exemplar for all of the other 'prime-containing' columns.

Now, $\pi(35) - \pi(5) = 11 - 3 = 8$, and $\pi(37) - \pi(5) = 13 - 3 = 9$. It then follows that **37** is prime.

$\pi(65) - \pi(35) = 18 - 11 = 7$, and $\pi(67) - \pi(35) = 19 - 11 = 8$, hence **67** is prime.

$\pi(95) - \pi(65) = 24 - 18 = 6$, and $\pi(97) - \pi(65) = 25 - 18 = 7$, hence **97** is prime.

$\pi(125) - \pi(95) = 30 - 24 = 6$, and $\pi(127) - \pi(95) = 31 - 24 = 7$, hence **127** is prime.

$\pi(155) - \pi(125) = 36 - 30 = 6$, and $\pi(157) - \pi(125) = 37 - 30 = 7$, hence **157** is prime.

$\pi(185) - \pi(155) = 42 - 36 = 6$, and $\pi(187) - \pi(155) = 42 - 36 = 6$, hence **187** is composite.

$\pi(215) - \pi(185) = 47 - 42 = 5$, and $\pi(217) - \pi(185) = 47 - 42 = 5$, hence **217** is composite.

$\pi(245) - \pi(215) = 53 - 47 = 6$, and $\pi(247) - \pi(215) = 53 - 47 = 6$, hence **247** is composite.

$\pi(275) - \pi(245) = 58 - 53 = 5$, **and** $\pi(277) - \pi(245) = 59 - 53 = 6$, hence **277** is prime.

Thus it is seen that we have reproduced the disposition of the primes in column 3 as far as 277.

Indeed, by using this method we can show that the next two numbers, **307** and **337** are prime and subsequently, the nature of any of the following numbers in the column.

It follows from the above that in obtaining the disposition of the primes in all the 'prime-containing' columns we can determine the gaps between the primes in the conventional 'linear' sense.

We may, of course, apply this procedure to any part of the matrix.

Consider, for example row 68 of the matrix shown on p8.

Let us assume pro tem that the status of the third element (**2017**) of this row has not been determined.

$\pi(2015) - \pi(1985) = 305 - 299 = 6$, and $\pi(2017) - \pi(1985) = 306 - 299 = 7$.

Hence the number **2017** is prime.

Of course, we can apply the procedure to any number in the matrix, e.g. the number **2063** lies in column 11 and row 69 of the matrix.

Now, $\pi(2061) - \pi(2015) = 310 - 305 = 5$, and $\pi(2063) - \pi(2015) = 311 - 305 = 6$.

Hence **2063** is prime.

Systematic application of this procedure will determine the nature of all of the elements of any row of the matrix.

For example consider row **83.**

Now, $\pi(2495) - \pi(2465) = 367 - 364 = 3$, hence there are **3** primes in this row.

Proceeding from the left, we have:

$\pi(2467) - \pi(2465) = 365 - 364 = 1$, hence **2467** is prime.

$\pi(2469) - \pi(2467) = 365 - 365 = 0$, hence **2469** is composite.

$\pi(2471) - \pi(2469) = 365 - 365 = 0$, hence **2471** is composite.

$\pi(2473) - \pi(2471) = 366 - 365 = 1$, hence **2473** is prime.

It is superfluous to examine **2475,** for it ends in **5** and cannot be prime.

$\pi(2477) - \pi(2475) = 367 - 366 = 1$, hence **2477** is prime.

We determined that there are **3** primes in row **83** and from the above we have exhausted the number of primes in that row. It then follows that all of the remaining numbers in the row are composite.

Using the same methods we can show that the number of primes in row **84** is **2,** and proceeding from the right that the last prime is **2503**. Hence, the gap between the last prime in row **83** and the first prime in row **84** ---none of the numbers between **2477** and **2503** is prime.

References

[1] The fluctuations of the prime counting function $\pi(x)$.

Compiled by A V Kulsha

www.primefan.ru/stuff/primes/table.html

[1a] On the determination of the primality of a number by the use of an accelerated version of trial division.

W M Fidler. May 2021.

ISBN 9783346493002.

[2] The first 50 million prime numbers

Don Zagier

Inaugural Lecture

Bonn University, May 1975.

[3] On the Number of Primes less than a Given Magnitude

G F B Riemann

The Monthly Notices of the Berlin Academy

November 1859.

[4] Ramanujan: Twelve Lectures on Subjects suggested by his life and work

G H Hardy

3rd Ed. New York: Chelsea, pp24-25

1999.

[5] Undersogelser angaende Maengden af Primtal under en given Graense

J P Gram

Kong. Dansk

Videnskab. Selsk. Skr (V1) 2, pp183-308 1884.

[6] Riemann Prime Number Counting Function—from Wolfram Mathworld.

mathworld.wolfram. com/RiemannPrimeCounting Function.html

[7] Computational strategies for the Riemann zeta function.

J M Borwein, D M Bradley, R E Crandall.

Journal of Computational and Applied Mathematics, 121(2000) 247 – 295.